MW00846652

SPACE SCIENCE

COMETS

BY BETSY RATHBURN

BELLWETHER MEDIA • MINNEAPOLIS, MN

TM

Are you ready to take it to the extreme? Torque books thrust you into the action-packed world of sports, vehicles, mystery, and adventure. These books may include dirt, smoke, fire, and chilling tales. **WARNING**: read at your own risk.

This edition first published in 2019 by Bellwether Media, Inc.

Library of Congress Cataloging-in-Publication Data

Names: Rathburn, Betsy, author.
Title: Comets / by Betsy Rathburn.
Description: Minneapolis, MN : Bellwether Media, Inc., [2019] | Series: Torque: Space Science | Audience: Grades 3 to 7. | Includes bibliographical references and index.
Identifiers: LCCN 2018001130 (print) | LCCN 2018010076 (ebook) | ISBN 9781681035994 (ebook) | ISBN 9781626178588 (hardcover : alk. paper)
Subjects: LCSH: Comets--Juvenile literature.
Classification: LCC QB721.5 (ebook) | LCC QB721.5 .R375 2019 (print) | DDC 523.6--dc23
LC record available at https://lccn.loc.gov/2018001130

Editor: Rebecca Sabelko Designer: Andrea Schneider

Printed in the United States of America, North Mankato, MN.

TABLE OF CONTENTS

A RARE SIGHT

It is April 1, 1997. The comet Hale-Bopp has just passed its **perihelion**. Behind the Sun, it is now the second-brightest object in the sky.

Hale-Bopp shines bright in the sky until December. People around the world look up to witness the rare sight. Hale-Bopp will not be back for more than 2,000 years!

HALE-BOPP

WHAT ARE COMETS?

From Earth, comets look like streaks of light across the sky. They bewildered humans for thousands of years.

Modern **telescopes** gave **astronomers** a closer look. Scientists discovered that comets look like oddly shaped, bumpy rocks. They are often covered in **craters**.

FUN FACT

AN EARLY SIGHTING

One of the earliest comet sightings was recorded in 240 BCE! Chinese astronomers observed Halley's Comet as it flew across the sky.

The three main parts of a comet are the **nucleus**, the **coma**, and the tail. The nucleus is a ball of ice, rock, and dust. Most are about 6 miles (10 kilometers) wide.

FUN FACT

KING-SIZED COMAS

A comet's coma can be thousands of miles wide.

Sometimes comets pass close to the Sun. This causes the ice to melt into **vapor**. Then, a cloud of dust and gas forms around the nucleus. This creates the coma.

Heat blows dust from the coma as the comet gets even closer to the Sun. This creates a long dust tail that can be seen from Earth!

THE PARTS OF A COMET

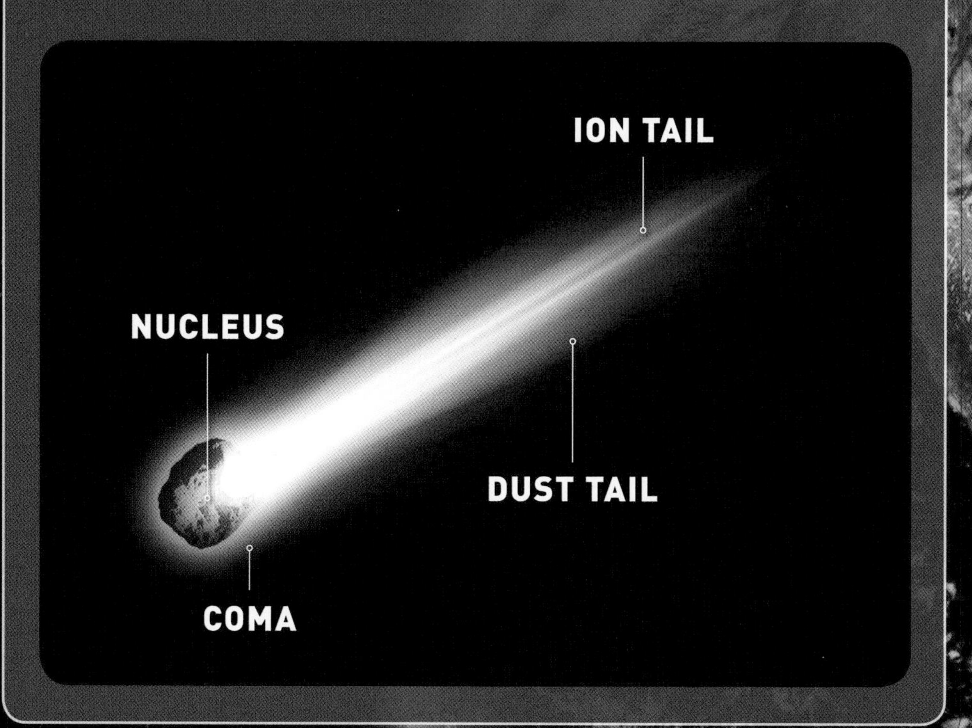

Comets also form a second, longer tail. This is called the **ion** tail. It is made of tiny gas **atoms**.

HOW DO COMETS FORM?

Comets were formed billions of years ago. They were left behind when a giant cloud of dust and gas **collapsed** to form the solar system.

Some of the **matter** joined to form the planets. But the rest became comets and other objects. Today, comets **orbit** the Sun.

WHERE ARE COMETS FOUND?

Astronomers have discovered over 5,000 comets in our solar system. But there may be billions more! Most have never been seen by humans.

Comets are found in the Kuiper Belt and the Oort Cloud. These cold, dark parts of space are millions of miles from Earth.

DWARF PLANET PLUTO
IN THE KUIPER BELT

Comets pass close to Earth as they orbit the Sun. Short-period comets likely come from the Kuiper Belt. Their orbit around the Sun takes less than 200 years.

HALLEY'S COMET

Type of comet: short-period
Named for: Edmund Halley
Length of orbit: about 76 years
Last time visible from Earth: 1986
Next time visible from Earth: 2061

Comets from the Oort Cloud take much longer to orbit the Sun. Some can only be seen from Earth every 2,000 years!

WHY DO WE STUDY COMETS?

Scientists use telescopes and spacecraft to study comets. They send **probes** thousands of miles through space to look at comets up close.

One famous probe was *Rosetta*. In 2014, it completed its ten-year journey to the comet 67P. Its pictures and studies gave scientists clues about how the solar system formed.

COMET 67P

ROSETTA PROBE

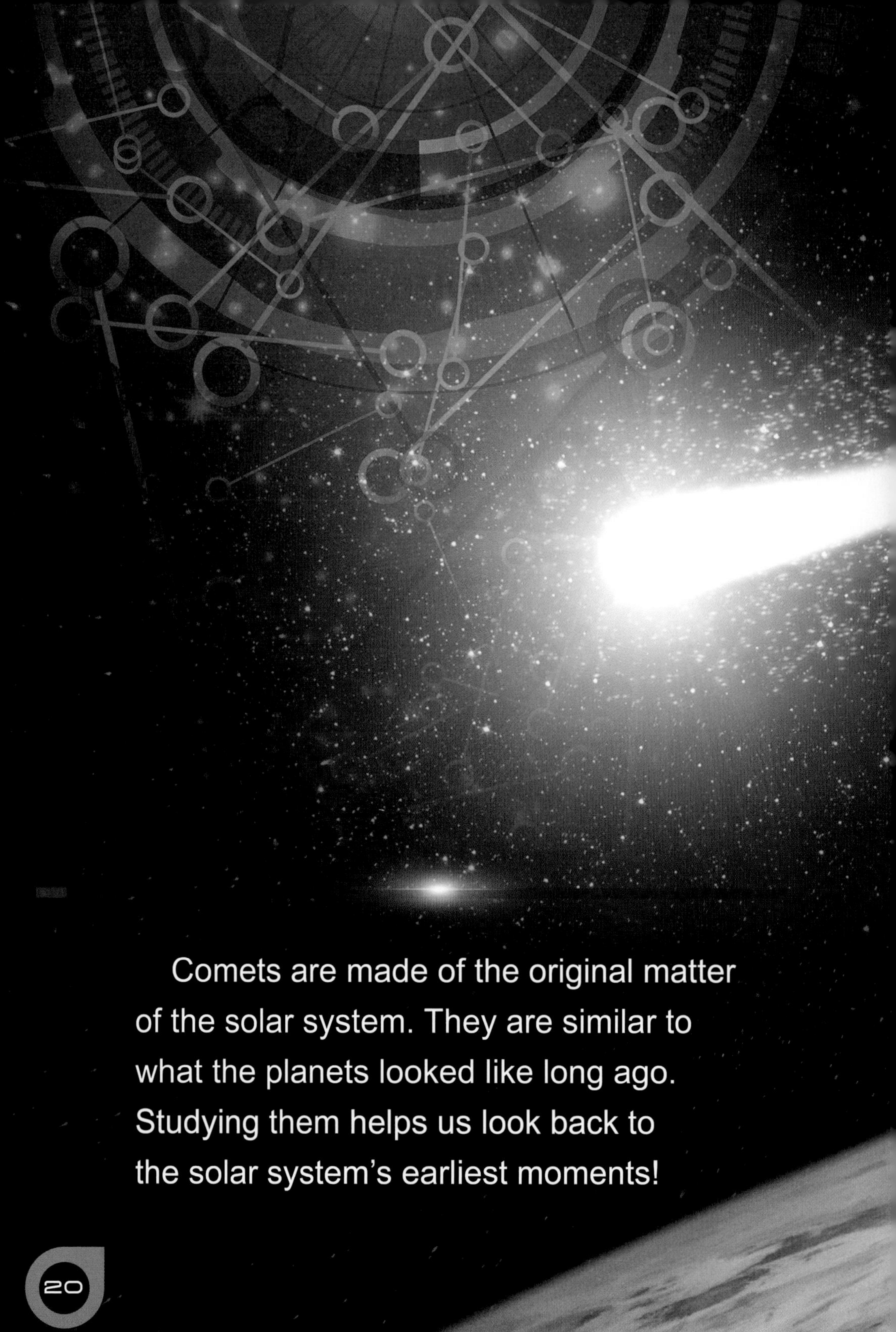

Comets are made of the original matter of the solar system. They are similar to what the planets looked like long ago. Studying them helps us look back to the solar system's earliest moments!

GLOSSARY

astronomers–people who study space

atoms–the smallest parts of materials

collapsed–caved in

coma–the cloud of gas that surrounds a comet's nucleus

craters–deep holes in the surface of a comet or other object

ion–an atom or group of atoms that has an electric charge

matter–the material something is made of

nucleus–the rocky part of a comet's head

orbit–to move around something in a fixed path

perihelion–the point nearest to the sun in a comet's orbit

probes–spacecraft designed to study faraway objects in space

telescopes–instruments used to view distant objects in outer space

vapor–gas

TO LEARN MORE

AT THE LIBRARY

Hamilton, John. *Rosetta: Voyage to a Comet*. Minneapolis, Minn.: Abdo Pub., 2017.

Hansen, Grace. *Comets*. Minneapolis, Minn.: Abdo Kids, 2018.

Knight, M.J. *Asteroids, Comets, and Meteors*. Tucson, Ariz.: Brown Bear Books, 2017.

ON THE WEB

Learning more about comets is as easy as 1, 2, 3.

1. Go to www.factsurfer.com

2. Enter "comets" into the search box.

3. Click the "Surf" button and you will see a list of related web sites.

With factsurfer.com, finding more information is just a click away.

INDEX

The images in this book are reproduced through the courtesy of: Milissa4like, front cover, pp. 3, 4, 6, 8, 10, 12, 14, 16, 18, 20, 23 (graphic); Temstock, front cover, p. 11 (comet); sdecoret, front cover (Earth); Alan Uster, front cover, pp. 2-3, 16 (Earth/Moon); Vadim Sadovski, pp. 2 (comets), 13 (Earth); ishift, p. 5 (inset); Stas Tolstnev, pp. 4-5; Yvonne Baur, pp. 6-7; TBStocker, pp. 8-9 (comet); Markus Gann, p. 9 (sun); solarseven, pp. 10, 14 (comet); Juergen Faelchle, p. 11 (Earth); DeoSum, p. 12; ikonacolor, pp. 12-13 (comets); NASA Images, p. 15 (Pluto/background); Paul Fleet, pp. 16-17 (comet); Kuiper Airborne Observatory/ NASA, p. 17 (inset); Elenarts, pp. 18-19 (Rosetta probe); Triff, pp. 18-19 (background); ESA/Rosetta/MPS for OSIRIS Team/ NASA, p. 19 (inset); Denis Tabler, pp. 20-21 (comet); Vladi333, pp. 20-21 (Earth).